AN ALPHABET IN SPACE

written and illustrated by

R. M. SMITH

CLARENCE-HENRY BOOKS

An Alphabet in Space
by R. M. Smith

Clarence-Henry Books • Alexandria, VA

Design and Layout by R. M. Smith

Summary: An ABCs space book that is designed to introduce nonfiction concepts to babies and toddlers.

ISBN-10: 098829091X
ISBN-13: 978-0988290914

First Edition
10 9 8 7 6 5 4 3 2 1

Printed and bound in U.S.A.

Way up in the sky,
most everything looks small,
but if you get a little closer,
it isn't small at all.

It's **rocky** and it moves really, really fast, as it **soars** easily and steadily past.

Aa is for Asteroid!

You might see it **shine** bright on a clear and cloudless **night**.

Bb is for Big Dipper!

It's **icy**, and it's **cold**. With a long **trail** of a **tail**, it's a sight to behold.

Cc is for Comet!

Where is it? What is it? Can you see it?
It's there, but it isn't.

Dd is for Dark Matter!

It rests at our **feet**, this special place,
this **rock** we call **home**,
floating in space.

Ee is for Earth!

Streaking the sky, they shine in the night, but they are gone in a blink, after once were so bright.

Ff is for Falling Stars!

Great in size, it's hard to see, there's one all around us, **spiraling** so free.

Gg is for Galaxy!

It helps us to **see** really **far**. It can **see** almost any planet and every star.

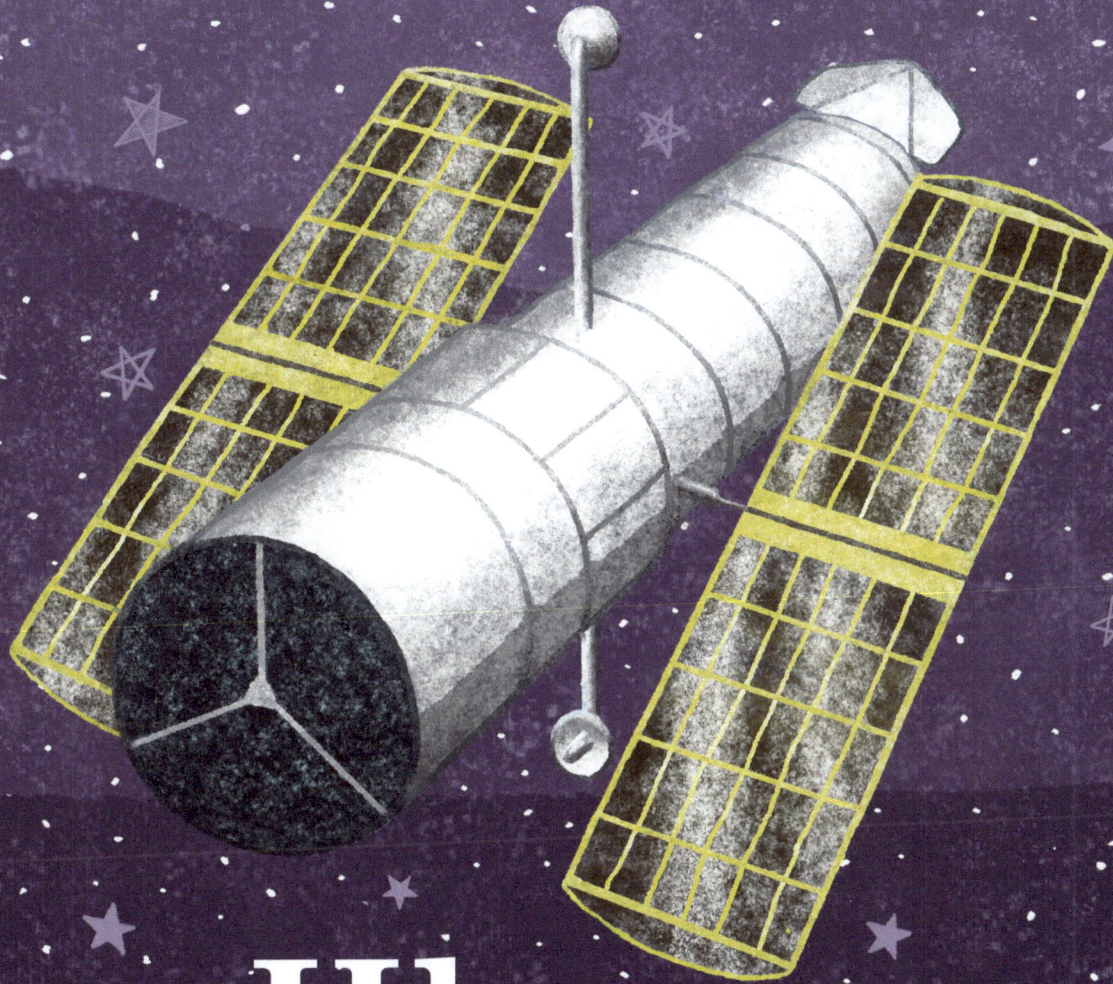

Hh is for Hubble Telescope!

People all over the **world** learn about space here. It **orbits** the earth, that much is clear.

Ii is for International Space Station!

It has a big red **spot**, right on its side,
this great big planet, so **tall** and **wide**.

Jj is for Jupiter!

Going in a big **circle**, every last one,
these dwarf planets and ice,
they **orbit** the sun.

Kk is for Kuiper Belt!

Like its bigger **sibling**,
but not quite as grand, you can see it
shine bright, far from our land.

Ll is for Little Dipper!

Our closest space rock, it's very **round**,
there are lots of **craters**
that can be found.

Mm is for Moon!

Very large and very **cold**, this planet is made up mostly of **liquid**, I'm told.

Nn is for Neptune!

Named for a Greek hunter from a time long ago, this group of stars **shimmer, sparkle**, and **glow.**

Oo is for Orion Constellation!

What is that **bursting ray** of colorful light? **On** and **off** it goes in the darkness of night.

Pp is for Pulsar!

Spinning round and round ever so **fast**, it **shoots** out light with a powerful **blast.**

Qq is for Quasar!

Blasting off into **space**,
it **carries** us and our dreams
to a brand new place.

Rr is for Rocket!

It's a planet with **rings**.
It's made up of many gases.
It's **bigger** than most things.

Ss is for Saturn!

It's a **bull** in space, far out and gruff,
it's got **horns** and **hooves**
and other animal stuff.

Tt is for Taurus Constellation!

What are these? We don't know.
Does it look like this? Maybe so.

Uu is for UFO!

To see it from afar, it looks like a **star**.
But the closer you see it
you'll know it's a **planet**.

Vv is for Venus!

A fading **star** losing its fuel. It still shines on but it's starting to **cool**.

Ww is for White Dwarf!

You **can't** see it at all. Nope! But you **can** if you're using an x-ray telescope.

Xx is for
X-Ray Telescope!

It's a star. We call it our **sun**. It's in our solar system. We have only **one**.

Yy is for Yellow Dwarf!

When you **jump** in space,
it'll put a **smile** on your face.

Zz is for Zero Gravity!

Solar System

Kuiper Belt

Uranus

Neptune

Jupiter

Mars

Venus

Saturn

Earth

Sun

Mercury